ALBERT EINSTEIN

FAILED IN THREE
CLASSICAL TESTS

GATOT SOEDARTO

Albert Einstein Failed In Three Classical Tests, Second Edition, 7 Sebtember 2017

ISBN-13: 978-1533381248

ISBN-10: 1533381240

Printed in USA by CreateSpace Independently Book Publisher

DEDICATION

I dedicated this book for all the people who love physics
and science

Contents

Preface

First of all, saying the classical tests of general relativity is just another way of saying the general relativity can be tested using the modern instruments we have today. Now we assumed if the general relativity had failed by the classical tests, why don't the general relativity just fail by the modern instruments?

Stephen Hawking tells us: "This proof of a German theory by British scientists was hailed as a great act of reconciliation between the two countries after the war. It is ionic, therefore, that later examination of the photographs taken on that expedition showed the errors were as great as the effect they were trying to measure. Their measurement had been sheer luck, or a case of knowing the result they wanted to get, not an uncommon occurrence in science. The light deflection has, however, been accurately confirmed by a number of later observations."

In the letter to the London Times on November 28, 1919, Albert Einstein described the theory of relativity and thanked his English colleagues for their understanding and testing of his work. He also mentioned three classical tests of general relativity: the perihelion precession of Mercury's orbit, the deflection of light by the Sun, and the gravitational redshift of light.

Einstein wrote: "The chief attraction of the theory lies in its logical completeness. If a single one of the conclusions drawn from it proves wrong, it must be given up; to modify it without destroying the whole structure seems to be impossible."

Until in the year 2010 I still believed that Einstein's general relativity is correct. I've published articles about Einstein's relativity in media; in the sense that I admired to Einstein, but I changed mind when I've read Lincoln Barnet's book 'The Universe and Dr. Einstein', London, 1949, foreword by Albert Einstein himself. I was surprised to read some paragraph in this book.

This book entitle Albert Einstein Failed in Three Classical Tests explains what's wrong with Einstein's hypothesis and the proving method of the general relativity. Furthermore, this book shows the incompatibility between special and general theory of relativity; and explains the three classical tests as requested by its founder; and discussed another tests which is done after Einstein died.

The most important thing Einstein had failed; it may be surprise you: Einstein had no idea on the basic of astronomy. In special and general theory of relativity Einstein had ignored refraction of light. He doesn't understand that space around the massive bodies is not empty vacuum, but the atmospheric medium. In the general relativity, Einstein's hypothesis and proposed test via eclipse weren't scientifically correct. He would like to measure the deflection of light by the Sun; but he proposed test to measure the deflection of light by Earth's atmosphere. He had not realized about that. Ironically testing via eclipse isn't scientifically correct and deeply wrong.

Einstein convinced to try to solve incompatibility between Special and General Relativity, until he died at the age of 76. Initially he seemed convinced that he could find a solution but gradually he became discouraged. He wrote to his old friend Maurice Solovine who had congratulated him for his seventieth birthday:"Now you think that I am looking back at my work with calm satisfaction. But on the closer look, it is quite different. There is not a single concept of which I am convinced that it will stand firm and I am not sure that I was on the right track after all"

Sebtember 7, 2017.   Gatot Soedarto

1. General Relativity Conflicts to Special Relativity

There are two kinds of Einstein's theory of relativity. The first theory is special relativity announced in the year 1905 with the famous equation of $E = mc^2$. Special theory of relativity describes the propagation of matter and light at high speed.

The second theory, general relativity, was announced in 1915. This theory was born stimulated by the new fact just realized later on by Einstein that his theory on Special Relativity was found to be inconsistent with the gravity theory of Newton, declaring that the space objects pull to each others with the force whose magnitude is determined by distance between the said objects.

His hypothesis on special theory of relativity concluded that the light velocity is the highest speed in this universe is in controversy with the gravity of Newton. The velocity of force attracting to each others among objects at the space, for example the attracting force of the moon causing the change in short time in the form of movement of sea tide on earth has the meaning that the gravity effect spreads at the boundless speed, not at the light velocity or lower.

Einstein begun by rejecting the ether theory

Lincoln Barnett tell us on the book 'The Universe and Dr.Einstein', London, 1949, page 38.

"Among those who pondered the enigma of the Michelson-Morley experiment was a young patent office examiner in Berne, named Albert Einstein. In 1905, when he was just twenty-six years old, he published a short paper suggesting an answer to the riddle in terms that opened up a new world of physical thought. He began by rejecting the ether theory and with it the whole idea of space as a fixed system or framework, absolutely at rest, within which it is possible to distinguish absolute from relative motion.

The one indisputable fact established by the Michelson-Morley experiment was that the velocity of light in unaffected by the motion of the earth. Einstein seized on this as a revelation of universal law. If the velocity of light is constant regardless of the earth's motion, he reasoned, it must be constant regardless of motion of any Sun, moon, star, meteor, or other syatem moving anywhere in universe. From this he drew a broader generalization, and asserted that the laws of nature are the same for alls uniformly moving system.

This simple statement is the essence of Einstein's Special theory of Relativity. It incorporates the Galilean Relativity Principle which states that mechanical laws are the same for all uniformly moving systems. But its phrasing is more comprehensive; for Einstein was thinking not only of mechanical laws laws but of the laws governing light and other electromagnetic phenomena. So he lumped them together in one fundamental postulate: all the phenomena of nature, all the law of nature, are the same for all systems that move uniformly relative to one another" [1]

Einstein's special theory of relativity states that the light velocity is the highest speed in this universe or nothing can travel faster than the speed of light. And the velocity of light is constant or light always travels at the same speed as 'a revelation of universal law'. Unfortunately, a revelation of universal law is incorrect.

"Nothing can travel faster than the speed of light."

"Light always travels at the same speed."

Have you heard these statements before? They are often quoted as results of Einstein's theory of relativity. Unfortunately, these statements are somewhat misleading. Let's add a few words to them to clarify. "Nothing can travel faster than the speed of light in a vacuum." "Light in a vacuum always travels at the same speed." Those additional three words in a vacuum are very important. A vacuum is a region with no matter in it. So a vacuum would not contain any dust particles. That's not to say that nothing ever travels faster than light. As light travels through different materials, it scatters off of the molecules in the material and is slowed down.The amount by which light slows in a given material is described by the index of refraction, n. The index of refraction of a material is defined by the speed of light in vacuum c divided by the speed of light through the material v: or $n = c/v$. [2]

Einstein admitted the ether theory

Ironically, in the year of 1920 Albert Einstein gave an address on 5 May at the University of Leiden; and Einstein admitted the ether theory.

Albert Einstein said: "Recapitulating, we may say that according to the general theory of relativity space is endowed with physical qualities; in this sense, therefore, there exists an ether. According to the general theory of relativity space without ether is unthinkable; for in such space there not only would be no propagation of light, but also no possibility of existence for standards of space and time."

Thus, Einstein's general relativity resolved conflicts between Newton theory of gravity and the special theory of relativity; but made conflicts with his special relativity.

Thought experiment proves special relativity is false

The most famously Einstein's experiment is the thought experiment of elevator. In his book 'The Universe and Dr.Einstein' renowned science writer Lincoln Barnett explains about Einstein's thought experiment as follows:

A cable is attached to the roof of the elevator; some supernatural force begins reeling in the cable; and the elevator travels "upward" with constant acceleration, i.e. progressively faster and faster. Again the men in the car have no idea where they are, and again they perform experiments to evaluate their situation. This time they notice that their feet press solidly against the floor come up beneath them.

If they release objects from their hands, the objects appear to "fall". If they toss object in a horizontal direction they do not move uniformly in a straight line, but describe a parabolic curve with respect to the floor.

And so the scientist, who have no idea that their windowless car actually is climbing through interstellar space, conclude that they are situated in quite ordinary circumstances in a stationary room rigidly attached to the earth and affected in normal measure by the force of gravity. There is no way for them to tell whether they are at rest in a gravitational field or ascending with constant acceleration through outer space where there is no gravity at all.

Figure 1.1: The objects appear to fall

From these fanciful occurrences Einstein drew a conclusion of great theoretical importance. To physicist it is known as the Principle of Equivalence of Gravitation and Inertia. It simply states that there is no way to distinguish the motion produced by inertial forces (acceleration, recoil, centrifugal force, etc) from motion produce by gravitational force.

The Einstein's equivalence principle is the heart and soul of gravitational theory, for it is possible to argue convincingly that if equivalence principle is valid, then gravitation must be a "curved spacetime" phenomenon, in other words, the effects of gravity must be equivalent to the effects of living in a curved spacetime.

But, from this fanciful experiment can also be concluded that the velocity of light is not constant and the velocity of light is not the same speed as the speed of light 'c' (300.000 Km/seconds). Let's look the experiment below.

Figure 1.2: The light beam comes from above the elevator.

In this experiment the elevator still travels upward with constant acceleration, and a light beam comes from above the elevator. If the observer within the elevator are equipped with sufficiently delicate instruments of measurement, they will be able to compute the speed of light beam. The results show that the speed of light beam is faster than the speed of light 'c' (because the elevator travels upward!).

The observer drew a conclusion of great theoretical importance that the speed of light is not constant or is not the same speed as the speed of light 'c': Thus, Einstein's special relativity really is false.

Einstein's thought experiment proves light travel in straight line

As before, the scene opens in an elevator ascending with constant acceleration through empty space, far from any gravitational field. A moment later as the car continues upward through space a beam of light is suddenly flashed through an aperture in the side of car. Since the velocity of light is a great, the beam traverses the distance between its points in a very small fraction of a second. Nevertheless, the car travels upward a certain distance in that interval, so the beam strikes the far wall a tiny fraction of inch below the point at which it entered.

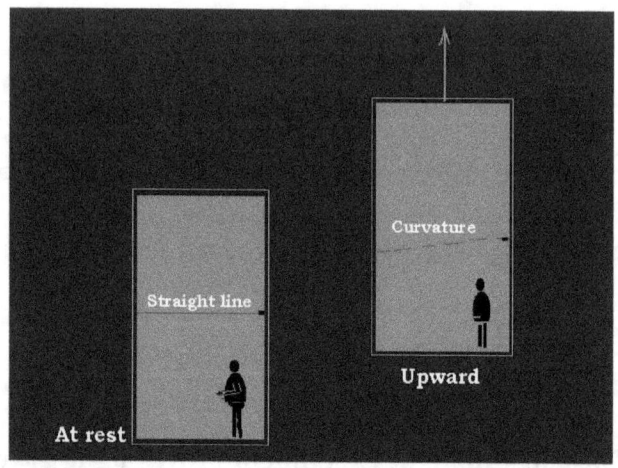

Figure 1.3: The curvature of light beam

If the observers within the car are equipped with sufficiently delicate instruments of measurement, they will be able to compute the curvature of the beam. But the question is, how will they explain it? They are still unaware of the motion of their car and believe themselves at rest in gravitational field. If the cling to Newtonian principles, they will be completely baffled because they will insist that light rays always travel in a straight line. But if they are familiar with the special theory of relativity they will remember that energy has mass in accordance with the equation m is E/C^2. Since light is a form of energy they will deduce that light has mass and will therefore be affected by a gravitational field. Hence the curvature of the beam.

From these purely theoretical considerations Einstein concluded that light, like any material object, travels in a curve when passing through the gravitational field of a massive body. He suggested that his theory could be put to test by observing the path of starlight in the gravitational field of the Sun. Since the stars are invisible by day, there is only one occasion when Sun and stars can be seen together in the sky, and that is during an eclipse.

Using Einstein's logic, we can draw conclusion another important consequence of the elevator thought experiment. Let's look the illustration below.

Figure 1.4: At rest, the light beam is directed slightly upward (white line).

Since the speed of elevator cannot be the same as the speed of light beam, in other words, it's very small when the speed is compared to the speed of light beam (more than one hundredth of the speed of light/300 million meters per second!). Moreover the light beam can be directed slightly upward (white line).

Then the observer inside the elevator see the light beam is not curve downward in a parabolic trajectory, but the light beam travels as a straight line (red line). The observer draw conclusion that light has no mass and travels in a straight line when passing through the gravitational field of a massive body. This experiment shows Newton's theory is correct.

So, we can draw conclusion that acceleration and gravitation cannot be equivalent. The bending of light cannot be produced by the Sun's gravitational field. It can be produced by the Earth's atmosphere, when the elevator travels upward within the Earth's atmosphere, not within the outer space.

In this case, using Einstein's thought experiment we can prove light travel in straight line; on other words Einstein's thought experiment proves Newton was right. What went wrong? Because the thought experiment can be used to obtain the result they wanted to get.

2. Einstein Had No Theoretical Reasons of Mercury's Orbit

Classical tests of general relativity theory

Actually Albert Einstein proposed three tests of the general theory of relativity:

1. the perihelion precession of Mercury's orbit.

2. the deflection of light by the Sun.

3. the gravitational redshift of light

In the letter to the London Times on November 28, 1919, he described the theory of relativity and thanked his English colleagues for their understanding and testing of his work. He also mentioned three classical tests with comments:

"The chief attraction of the theory lies in its logical completeness. If a single one of the conclusions drawn from it proves wrong, it must be given up; to modify it without destroying the whole structure seems to be impossible."

The Solar System Planets

A planetary system is a set of gravitationally bound non-stellar objects in orbit around a star or star system. The entire planet revolving around the Sun due the force of gravity between the Sun and the planet; they didn't to get collide or fall toward the Sun due to the fact, there is no motion without a cause. There are also forces from planets and medium between them in the opposite direction. These phenomena can be pictured in the formula:

Energy in =Energy out, and the result is a balance.

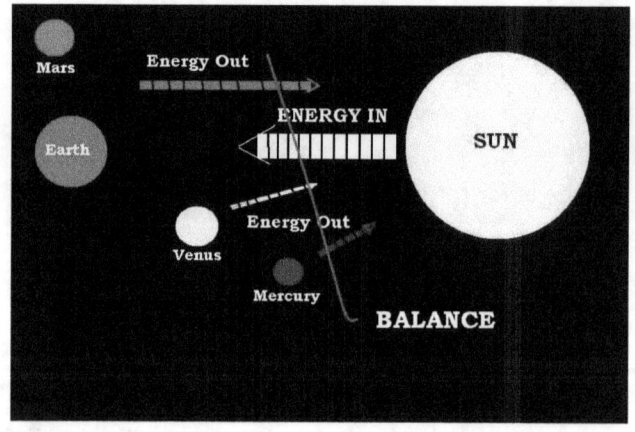

Figure 2.1: All things in Universe is energy, Energy In = Energy Out

Think about it a bit. It's an important formula. In fact, it's the central concept to all planetary system. Suppose, then, we know Newton's gravity predicted the planetary orbits so accurately, all planets follow an elliptical orbit, but should be taking into account deviation due to their distance from the Sun, and the gravitational pull of the other planets.

Why do the planets go around the Sun?

But now we still have the question of why anything orbits something else. The reasons are complicated but the first good explanation was provided by one of the greatest scientists ever, Isaac Newton, he is widely considered to be one of the most brilliant, important, and productive scientists ever to have lived.[3]

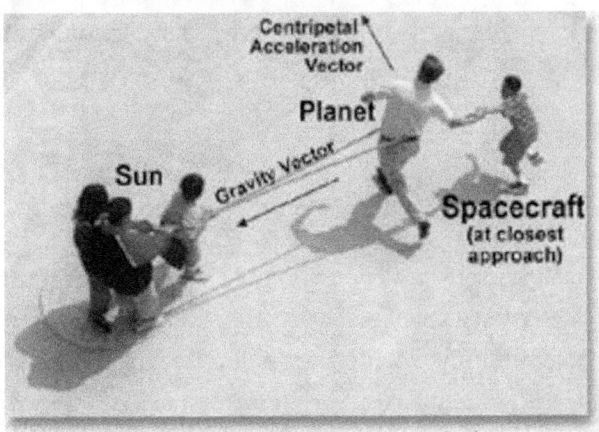

This activity demonstrates not only how a planet can orbit the Sun, but also give a boost in speed to a passing spacecraft.

Figure 2.2: The planets go around the Sun

These activity demonstrates not only how a planet can orbit the Sun, but also give a boost in speed to a passing spacecraft.

Elliptic orbit

In astrodynamics or celestial mechanics, an elliptic orbit or elliptical orbit is a Kepler orbit with an eccentricity of less than 1; this includes the special case of a circular orbit, with eccentricity equal to 0. In a stricter sense, it is a Kepler orbit with the eccentricity greater than 0 and less than 1 (thus excluding the circular orbit). In a wider sense, it is a Kepler orbit with negative energy. This includes the radial elliptic orbit, with eccentricity equal to 1.

Mercury's orbit is the most eccentric

Scientists classify orbits by the shape that they trace through space. Scientists use the term "eccentricity" to explain how round or oblong the orbit is. The higher the eccentricity, the more "squished" the orbit appears.

Most astronomical objects orbit some body that it is more massive than it is. For example, the moon orbits the Earth, the Earth orbits the Sun, and the

Sun orbits the galactic center. Each of these orbits take the form of an ellipse. Because these bodies do not travel in a perfect circle, they are not always the same distance from the center of their orbit or the object that they orbit. When an object is as close as it gets to the object it is orbiting, it is said to be at perihelion. By contrast, the farthest point of the ellipse from the body being orbited is called the aphelion. The orbiting object travels the fastest while it is close to the perihelion and slowest when it is at the aphelion.

Mercury speeds around the sun every 88 Earth days, traveling through space at nearly 112,000 mph (180,000 km/h), faster than any other planet. Its oval-shaped orbit is highly elliptical, taking Mercury as close as 29 million miles(47 million km) and as far as 43 million miles (70 million km) from the sun. Mercury's orbit experiences some of the most bizarre conditions.

Perihelion and Aphelion data:

Mercury

Perihelion is about 46,000,000 Km from the Sun, and Aphelion is about 69,800,000 Km.

Venus

Perihelion Venus is about 107,476,259 Km from the Sun, and Aphelion is about 108,942,109 Km.

Earth

Perihelion is about 147,100,000 Km from the Sun, and Aphelion is about 152,100,000 Km,

Mars

Perihelion is about 206,655,215 Km from the Sun and Aphelion is about 249,232,432 Km.

Wikipedia tells us, Newton derived an early theorem which attempted to explain apsidal precession. This theorem is historically notable, but it was never widely used and it proposed forces which have been found not to exist, making the theorem invalid. This theorem of revolving orbits remained largely unknown and undeveloped for over three centuries.

But, now we can explain the Sun's energy fluctuation using Newton's theory of gravity, not Einstein's theory of gravity.

The Sun's fluctuations

The study in the year 2010 informs us that the Sun's energy can rise and falls. The Sun's fluctuations caused partial collapse of Earth's atmosphere. From this study, triangle's energy concept predicted that the Sun's energy fluctuation as the cause of the unusual of Mercury's orbit. In other words, the Sun's energy fluctuation caused perihelion of Mercury doesn't happen at the same place but moves slowly around the Sun.

The point of closest approach of Mercury to the sun does not always occur at the same place, but moves around the Sun. It is called a precession of the perihelion of Mercury.

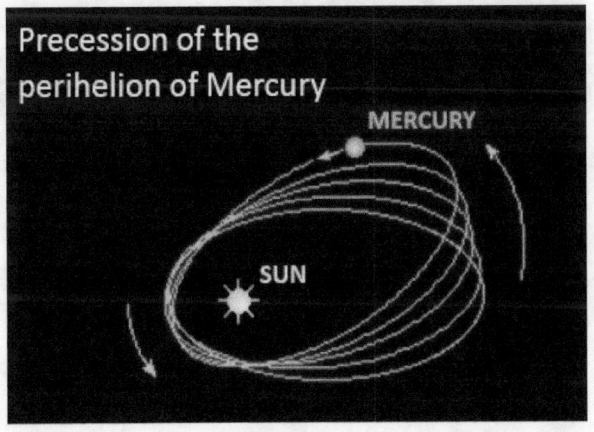

Figure 2.3: A precession of the perihelion of Mercury.

Let's imagine Sun's energy fluctuation such as the occurrence of waves in the ocean. Sea waves can rise and falls. Sometime sea waves will generate a kind of "explosion", when the water mass collides and generate splashes at surface of the ocean. It goes on continuously and makes fluctuation of sea waves energy.

Figure 2.4: Sea waves splashes landscape ocean, just like the Sun's energy fluctuation-Image WallpaperUp

Now, what cause the Mercury's orbit is the most eccentric? In the sense that the perihelion of Mercury doesn't happen at the same place but moves slowly around the Sun.

Mercury is very close to the Sun and the gravitational pull of the Sun is very high; but Mercury doesn't fall towards the Sun because there are some forces from other planets and medium between them in opposite direction. The Sun's fluctuation energy have an impact and makes the Mercury's orbit is the most eccentric.

Thus, a precession of the perihelion of Mercury can be explained using Newton's theory of gravity; in the sense that gravity is a force. Therefore, a precession of the perihelion of Mercury can not be explained using Einstein's theory of gravity; in the sense that gravity is nothing about force but a warping of spacetime. The idea of a warping of spacetime has no scientific reasoning to explain the fluctuation energy as the cause of Mercury orbits.

The initial location of Mercury before drifting towards the Sun under the influence of the spinning-gravitational effect of the Sun would determine how eccentric its orbit in accordance with its mass content. Orbits of all planets in solar system are eccentric even if the Earth's orbit is not completely circular. As a matter of fact the Earth's orbit is slightly eccentric. Then, it is not a big surprise that Mercury's orbit is so eccentric, that's because Mercury is the closest planet to the Sun.

When we compared to other distant planets, solar energy fluctuations have a major effect on Mercury. From this, it can be predicted that the planets furthest from the Sun, Neptune and dwarf planet Pluto; their orbits should be close to completely circular.

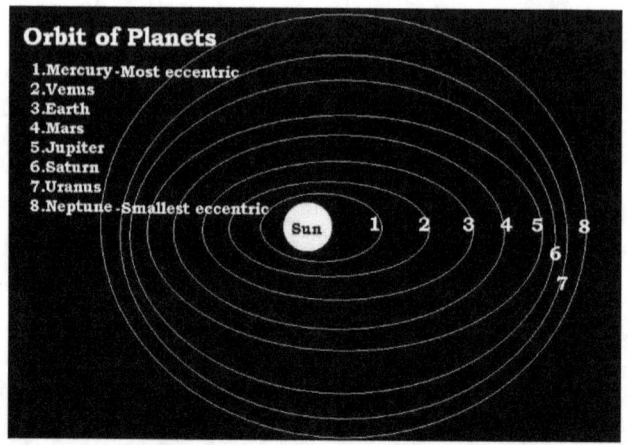

Figure 2.5: **Neptune is the smallest eccentric.**

This prediction matched with Neptune's orbit, but may be doesn't match with Pluto, for example, if we taking account the effect of gravitational pull from the other planets and Pluto's mass.

Neptune

Perihelion is about 4444.5 million Km from the Sun, and Aphelion is about 4545.7 million Km.

In general relativity, this remaining precession, or change of orientation of the orbital ellipse within its orbital plane, is explained by gravitation being mediated by the curvature of spacetime. Einstein showed that general relativity agrees closely with the observed amount of perihelion shift. This was a powerful factor motivating the adoption of general relativity.

The problem of Einstein's gravity; is not an ordinary force, but rather a property of space-time geometry, how can predict precession of the perihelion of Mercury more accurate than Newton?

Let's look at curved space according to Einstein's general theory of relativity

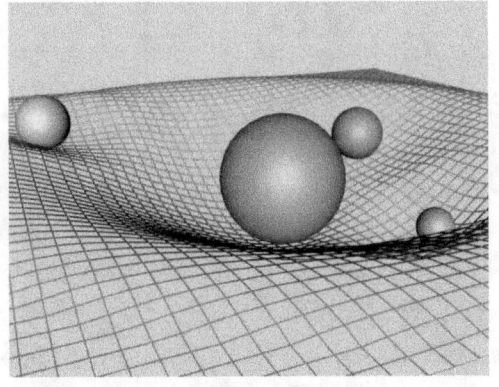

Figure 2.6: The curvature of space due to gravity
(UniverseToday)

That's why without gravity as a real force as Newton's gravity, the Mercury orbit. can't be calculated Therefore, Einstein's prediction of the Mercury orbit needs to be questioned.

General relativity's prediction; of an extra 43" per century, while having no mathematical or theoretical reasons.

Einstein's Mercury orbit was challenged by several scientists including Dr. Thomas Van Flandern astronomer who worked at the U.S. Naval Observatory in Washington. Thomas Van Flandern asked a colleague at the University of Maryland, who as a young man had overlapped with Einstein at Princeton's Institute for Advanced Study, how in his opinion Einstein had arrived at the correct multiplier. This man said it was his impression that, "knowing the answer," Einstein had "jiggered the arguments until they came out with the right value."[4]

Another scientist said:"He simply assumed the period of precession, since his number matched historical equations. As I show, this assumption was false, since his number .45 applied to the curvature of his field at the distance of Mercury's orbit. That is, it was a constant, applying during one second or one century. He needed more math in order to apply that curvature to the precession problem, but he never did that math. He simply applied the curvature number directly to precession. This is not only mathematically disallowed, it is gloriously negligent. I don't know— and probably no one knows or ever did know— whether this was an oversight or a purposeful fudge. It may be that he couldn't see how to get from the curvature to the precession, so he just took what he had and ran with it. Because his audience was already monumentally confused, no one noticed."[5]

Einstein had no mathematical and theoretical reasons on Mercury's orbit

Mathematics in physics actually is about a way of modeling patterns that we see in the real world. It is similar with thought experiment, a mathemathic equation can be made to obtain the result they wanted to get.

Is the mathematic true and then the theory should be true? No, for example, according to Maxwell's equations, it's possible to have a "magnetic monopole"—a magnet with a North pole but no South pole. Magnetic monopoles have never been discovered, and many scientists believe they don't really exist.

It is impossible to make magnetic monopoles from a bar magnet. If a bar magnet is cut in half, it is not the case that one half has the north pole and the other half has the south pole. Instead, each piece has its own north and south poles. A magnetic monopole cannot be created from normal matter such as atoms and electrons, but would instead be a new elementary particle.

If you study Einstein Field Equation of gravitation (EFE), you'll find EFE is based on assumptions, thought experiment, and Riemann geometry. In the case of Einstein's thought experiments, we know that there are full of logical fallacies; a kind of a false equivalence and the fallacy of composition.

One of the basic topics in Riemannian geometry is the study of curved surfaces in general. Riemann geometry also study higher dimensional spaces. But, there are no practical applications of Riemann geometry in astronomy. Riemann did not take an interest in the space of the astronomers. Questions about the global properties of space he cut short as "idle questions.".

Projecting a sphere to a plane is very important to make a map used in astronomy. But, the astronomy doesn't using Riemann geometry.

Riemann's ambition was nothing less than a total theory of physics, including electricity, magnetism, gravity and light, based on a single mathematical law. In this case Riemann had failed. Riemann's ambition was forwarded by Einstein; and Einstein had also failed to find the Unified Field Theory.

The Einstein Field Equations of Gravitation:

$$ R_{\mu\nu} - \frac{1}{2} R \, g_{\mu\nu} + \Lambda \, g_{\mu\nu} = \frac{8\pi G}{c^4} \, T_{\mu\nu} $$

The left-hand side of that equation is a matrix of numbers (curvature of spacetime). Mathematically, spacetime is a manifold. The right-hand side is a matrix of quantum operators, each of which has an expectation value. This, at some level, makes no sense.

To get his equation, among other Einstein assumes that 'in spaces where matter is absent'; this assumption is reasonless to reality. Outer space, or just space, is the void that exists between celestial bodies, including Earth. It is not completely empty, but consists of a hard vacuum containing a low density of particles, predominantly plasma of hydrogen and helium as well as electromagnetic radiation, magnetic fields, neutrinos, dust, and cosmic rays.

Mercury is very close to the Sun and the gravitational pull of the Sun is very high. As a manifold in Riemann geometry, spacetime can't explains why doesn't Mercury fall towards the Sun? And how did spacetime makes the Mercury's orbit is the most eccentric?

The main cause of the most eccentric orbit of Mercury is the Sun's fluctuation energy. This can not be explained by Einstein's theory of gravity, but can only be explained by Newton's theory of gravity.

The most popular example to explains Einstein's

gravity is a rubber sheet and a ball as it shows on YouTube and on many of Websites. Einstein's gravity contains nothing about force. It describes the behavior of objects in a gravitational field not in terms of 'attraction 'but simply in terms of the paths they follow. To Einstein, gravitation is simple part of inertia; the movement of the stars and the planets arise from their inherent inertia; and the courses they follow are determined by the metric properties of space or a warping of spacetime.

A rubber sheet warped by a heavy stone is an excellent analogy. A rubber sheet as seen on the figure below illustrated a warping of spacetime.

Figure 2.6: A warping of spacetime (**Image from universe today**)

Put and push a ball on the rubber sheet, then the

ball will move around the hole. But, within 2–3 seconds the ball will go into the hole. Of course, a ball unable to move again.

Teaching using a rubber sheet actually is misleading. It should be explained that the ball will continue to move circular around a hole -a warping spacetime-because the ball has an orbital velocity. It means, there is a force. Einstein said, the movement of the stars and the planets arise from their inherent inertia.

We know that inertia is one of the primary manifestations of mass, which is a quantitative property of physical systems. Isaac Newton defined inertia as his first law in his Philosophiæ Naturalis Principia Mathematica, which states:

"The vis insita, or innate force of matter, is a power of resisting by which every body, as much as in it lies, endeavours to preserve its present state, whether it be of rest or of moving uniformly forward in a straight line."

Therefore, Einstein's idea 'gravity is nothing about force' is inconsistent, and in fact, Einstein's theory of gravity is very dependent on Newton's theory. Thus, it clear that, "knowing the answer," Einstein had "jiggered the arguments until they came out with the right value." are correct.

3. Deflection of Light: Einstein Had No Idea on the Basic of Astronomy

Hypothesis and Einstein proposed test of general relativity via eclipse are closely related to astronomy, especially celestial navigation. For understanding the hypothesis and the test via eclipse; both are correct or incorrect, physics training is needed; but more importantly is celestial navigation training.

One of the fundamental concepts in astronomy, when taking observe for finding the angle of bending of star light in the sky, knowing the altitude of star is just as important as measuring the time of moving celestial bodies. We must know the altitude of star quite exactly if observations are to be of any use. The reason for this is that the entire of calculations depending on the altitude of star.

Gravitational deflection of light

The first calculation of the deflection of light by mass was published by the German astronomer Johann Georg von Soldner in 1801. Soldner showed that rays from a distant star skimming the Sun's surface would be deflected through an angle of about 0.9 seconds of arc, or one quarter of a thousandth of a degree. This angle corresponds to the apparent diameter of a compact disc (CD) viewed from a distance of about 30 kilometers (nearly 20 miles). Soldner's calculations were based on Newton's laws of motion and gravitation, and the assumption that light behaves like very fast moving particles. As far as we know, neither Soldner nor later astronomers attempted to verify this prediction, and for good reason: Such an attempt would have been far beyond the capability of early 19th century astronomical instruments.

Light deflection in general relativity. Over a century later, in the early 20th century, Einstein developed his theory of general relativity. Einstein calculated that the deflection predicted by his theory would be twice the Newtonian value.

The following image shows the deflection of light rays that pass close to a spherical mass. To make the effect visible, this mass was chosen to have the same value as the Sun's but to have a diameter five thousand times smaller (i.e., a density 125 billion times larger) than the Sun's.[6]

It is interesting to note, first, on the einstein-online that says 'Einstein calculated that the deflection predicted by his theory would be twice the Newtonian value.' Second, on the sciencenews.org that explains the bending of light as seen from Earth; shift angle 1.75 seconds of arc.[7]

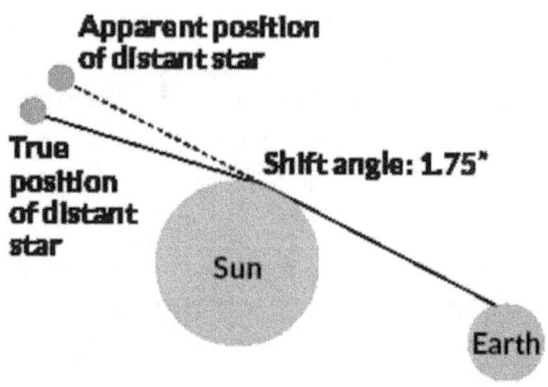

Figure 3.1: Gravitational deflection of light

First, Newton himself never predicts deflection of light by the Sun; to be more precisely, absence of evidence about it. Newton never mentioned the bending of light is about 0.9 seconds of arc.

A paper of Domingos S.L.Soares, 'Newtonian gravitational deflection of light revisited', submitted on 3 Aug 2005; describes the angle of deflection of a light ray by the gravitational field of the Sun, at grazing incidence, is calculated by strict and straightforward classical Newtonian means using the corpuscular model of light. The calculation is presented in the historical and scientific contexts of Newton's and of modern views of the problem.

Domingos S.L.Soares drew a conlusion, there is still the question of why Newton did not discuss the possibility of light ray deflection by a massive heavenly body. Of course, he was well acquainted with the relevant astronomical observations.Solar eclipses were certainly of his knowledge and could certainly motivate digressions on the gravitational bending of light.

Second, general relativity also predicts that gravity should bend light, but for very different reasons. In general relativity, a massive object distorts spacetime itself, and light simply takes the straightest path. You have to work through the numbers, but if you do, you discover that this means light bends twice as much in general relativity as in Newtonian gravity.

According to Einstein, the star light visible around the sun would be bent inwards, toward the sun at the time when passing through the gravity field of the sun. Einstein calculated the level of their deviation and predicted that for the stars observed being the closest to the Sun, their deviation was about 1.75 seconds of arc-See Figure 3.1.

For more than 100 years all physicists and astrophysicists are very familiar with the illustration in Figure 3.1 above; but, did they realize that the above illustration has no meaning whatsoever?

The above illustration shows Einstein had been

failed to understand the basic of astronomy.

What is the reason? This prediction is not meaningful in scientifically of astronomy when it is not explained the altitude of the star. The important things to be noted; the amount of 1.75 seconds of arc without taking into account the altitude of the star or the Sun during eclipse as the object of observation. This is a fatal mistake; because the deviation of starlight will always varies depending on the altitude of the object of observation.

In the view of scientifically of astronomy the illustration on the Figure 3.1 is not gravitational deflection of light, but deflection of light by Earth's atmosphere. The illustration explains the observations of stars during a solar eclipse, and the observations are made using a telescope; therefore, the result is always deflection of light by Earth's atmosphere, not by gravity.

From the above discussion we can draw a conclusion that Einstein had no idea on the basic of astronomy. To be more specific it can be explained as follow:

"From these purely theoretical considerations Einstein concluded that light, like any material object, travels in a curve when passing through the gravitational field of a massive body. He suggested that his theory could be put to test by observing the path of starlight in the gravitational field of the Sun. Since the stars are invisible by day, there is only one occasion when Sun and stars can be seen together in the sky, and that is during an eclipse.

Einstein proposed therefore, that photographs be taken of the stars immediately bordering the darkened face of the sun during an eclipse and compared with photographs of those same stars made at another time. According to his theory, the light from the stars surrounding the Sun should be bent inward, toward the Sun, in traversing the Sun's gravitational field; hence the images of these stars should appear to observer on earth to be shifted outward from their usual positions in the sky.

Einstein calculated the degree of deflection that should be observed and predicted that for the stars closest to the Sun the deviation would be about 1.75 seconds of an arc.Since he staked his whole General Theory of Relativity on this test, men of science throughout the world anxiously awaited the findings of expeditions which journeyed to equatorial regions to photograph the eclipse of May 29, 1919. When their pictures were developed and examined, the deflection of the starlight in the gravitational field of the sun was found to average 1.64 seconds—a figure as close to perfect agreement with Einstein's prediction as the accuracy of instruments allowed."(Lincoln Barnett, The Universe and

Dr.Einstein, page 78)

Explanation:

1.Einstein's prediction of 1.75 sec.arc without taking into account the altitude of the star/Sun as the object of observation. This is a fatal mistake because this prediction has no scientific meaning in astronomy. The value bending of starlight isn't the same at the minimum solar eclipse and at the maximum solar eclipse. The important thing to be note, in astronomy, deviation or bending of starlight will always vary depending on the altitude of the object of observation. In this case, hypothesis Einstein is not valid. It means Einstein hypothesis of general relativity doesn't meet requirements of scientific method.

2."Einstein proposed therefore, that photographs be taken of the stars immediately bordering the darkened face of the sun during an eclipse and compared with photographs of those same stars made at another time. According to his theory, the light from the stars surrounding the Sun should be bent inward, toward the Sun, in traversing the Sun's gravitational field; hence the images of these stars should appear to observer on earth to be shifted outward from their usual positions in the sky."

This sub-paragraph shows that he wants measuring deflection of light by the Sun; but he proposed test measuring deflection of light by Earth's atmosphere; he had not realized about that. Ironically, this test is not scientifically correct and deeply wrong.

a. When observation is made from Earth, not from outer space, the result is always deflection of starlight by Earth's atmosphere.

b. Calculating the angle of deflection of light (bending, deviation), in astronomy applies direct observation and instantaneous. "compared with photographs of those same stars made at another time" is not scietifically correct

In 'A Brief History of Time', Stephen Hawking write:

"This proof of a German theory by British scientists was hailed as a great act of reconciliation between the two countries after the war. It is ionic, therefore, that later examination of the photographs taken on that expedition showed the errors were as great as the effect they were trying to measure. Their measurement had been sheer luck, or a case of knowing the result they wanted to get, not an uncommon occurrence in science. The light deflection has, however, been accurately confirmed by a number of later observations."

The later eclipse observations was conducted 6 times; as we know the 1919 eclipse experiment was error, but repeated in the year of 1922, 1929, 1936, 1947, 1952, and in the year of 1973. The result of the entire eclipse observations were declared Einstein's general relativity was right. Here a big question: repeating the wrong method for decades, how could it happen in modern science?

Astronomical data of 1919-1973 eclipse experiments

1.Astronomical data of 1919 solar eclipse

We wiil easily able to see in the Nautical Almanac of 1919 the deviation of a certain star that closest to the Sun during maximum solar eclipse. The apparent altitude of maximum eclipse in West Africa is about 70,6 degrees, and the astronomical refraction is about 21 sec.arc, it more than 10 times greater than Einstein's prediction.

Event	Date	Time (UT)	Alt	Azi
1919 Eclipse - West Africa Lat : 2.1089^0 N Long: 16.875^0 W				
Start of partial eclipse	1919/05/29	11.31.05.0	60.1^0	047.8^0
Maximum eclipse	1919/05/29	13.07.10.1	70.6^0	358.2^0
End of partial eclipse	1919/05/29	14.44.19.3	58.9^0	310.5^0

Deviation of a starlight closest to the Sun (App.Lat Corr for 70.6^0)

Astronomical refraction   DIP/Terrestrial refraction

- 0,35 minutes = - 21 sec.arc   - 2,8 minutes = - 168 sec.arc

Figure 3.2: 1919 eclipse data of West Africa

Moreover, if taking into account the deviation caused by terrestrial refraction which is value depend on elevation of the place of observation, and always greater than astronomical refraction, as shown on the Figure 3.2; deviation is about 189 sec.arc, it's more than 100 times greater than Einstein's prediction. The terrestrial refraction/DIP value is calculated with the minimum height of eye 2, 5 meters.

On the Figure 3.3 below we can see the astronomical data of 1919 eclipse in Brazil. The altitude of maximum eclipse is about 21.7 degrees and astronomical refraction is about 153 sec.arc. It's more than 7 times greater than the value in West-Africa. DIP or terrestrial refraction is about 168 sec.arc.

1919 Eclipse - Brazil
Lat : 11.5239⁰ S
Long: 56.0742⁰ W

Event	Date	Time (UT)	Alt	Azi
Start of partial eclipse	1919/05/29	10.34.24.1	07.8⁰	1⁰
Maximum eclipse	1919/05/29	11.37.42.1	21.7⁰	061.1⁰
End of total eclipse	1919/05/29	11.39.38.9	22.1⁰	060.9⁰

Deviation of a starlight closest to the Sun (App.Lat Corr for 21.7⁰)

Astronomical refraction	DIP/Terrestrial refraction
- 2,55 minutes = - 153 sec.arc	- 2,8 minutes = - 168 sec.arc

Figure 3.3: 1919 eclipse data of Brazil

The table on the Figure 3.3 shows the invalidity of Einstein's prediction of 1.75 sec.arc without taking into account the altitude of the star/Sun as the object of observation.

2.Astronomical data of 1922 eclipse in Australia.

The altitude of maximum eclipse in Australia is about 24.7 degrees and astronomical refraction is about -123 sec.arc. DIP/terrestrial refraction is about -168 sec arc as shown on the Figure 3.4 below.

1922 Eclipse - Australia

Lat : 30.1451^0 S
Long: 146.25^0 E

Event	Date	Time (UT)	Alt	Azi
Start of partial eclipse	1922/09/21	05.00.12.4	38.6^0	299.3^0
Maximum eclipse	1922/09/21	06.10.00.3	24.7^0	286.8^0
End of total eclipse	1922/09/21	07.12.50.4	11.5^0	277.9^0

Deviation of a starlight closest to the Sun (App.Lat Corr for 24.7^0)

Astronomical refraction	DIP/Terrestrial refraction
- 2,05 minutes = -123 sec.arc	- 2,8 minutes = -168 sec.arc.

Figure 3.4: 1922 eclipse data of Australia

It's means deflection of light is about -291 sec arc. The value of deflection of light is more than 100 times greater than Einstein's prediction. Thus, that's proves Einstein's prediction really doesn't work.

3.Astronomical data of 1929 solar eclipse in Sumatra

The altitude of maximum 1929 eclipse in Sumatra is about 71.6 degrees, and the astronomical refraction is about -21 sec arc. The DIP or terrestrial refraction is about -168 sec arc. See on the Figure 3.5 below.

1929 Eclipse - Sumatra

Lat : 0.7031⁰ N
Long: 94.2188⁰ E

Event	Date	Time (UT)	Alt	Azi
Start of partial eclipse	1929/05/09	04.36.25.0	65.6⁰	043.1⁰
Maximum eclipse	1929/05/09	06.07.00.4	71.6⁰	343.2⁰
End of total eclipse	1929/05/09	07.40.21.2	56.6⁰	303.2⁰

Deviation of a starlight closest to the Sun (App.Lat Corr for 71.6⁰)

Astronomical refraction	DIP/Terrestrial refraction
- 0,35 minutes = - 21 sec.arc.	- 2,8 minutes = - 168 sec.arc.

Figure 3.5: 1929 eclipse data of Sumatra

Its means the value deflection of light is about -189 sec arc, that's more than 100 times greater than Einstein's prediction. With all due respect we must say again: "Einstein's prediction really doesn't work."

4.Astronomical data of 1952 solar eclipse in Sudan (Africa)

The altitude of maximum 1952 eclipse in Sudan is about 61.8 degrees. The astronomical refraction is about -33 sec arc, and the terrestrial refration /DIP is about -168 sec arc. See on the table 3.6 below.

1952 Eclipse - Sudan (Africa)

Lat : 14.5064^0 N
Long: 31.5571^0 E

Event	Date	Time (UT)	Alt	Azi
Start of partial eclipse	1952/02/25	07.40.43.3	46.5^0	121.3^0
Maximum eclipse	1952/02/25	09.06.43.0	61.8^0	147.1^0
End of total eclipse	1952/02/25	10.34.24.1	65.2^0	196.2^0

Deviation of a starlight closest to the Sun (App.Lat Corr for 61.8^0)
Astronomical refraction   DIP/Terrestrial refraction
- 0,55 minutes = - 33 sec.arc.   - 2,8 minutes = - 168 sec.arc.

Figure 3.6: 1952 eclipse data of Sudan

Its means the value of deflection of light is about - 201 sec arc, and again that's more than 100 times greater than Einstein's prediction.

The observers in Sudan may be being able to see the North Star of Polaris during the maximum solar eclipse. Altitude of Polaris is the same with latitude of Sudan; it about 14.5 degrees. That's actually the best opportunity to re-examine the validity of the curvature of spacetime.

General relativity predicts that the light track of star being closest to the Sun would be deflected 1.75 sec arc. If the light star is far away to the Sun would be deflected smaller or may be negligible difference between apparent position and actual position of star. In this case, the Polaris star is far away from the Sun during solar eclipse. So we can calculate to find the value of the deflection of Polaris using the Nautical Almanac.

From the Nautical Almanac of 1919 we found the astronomical refraction of Polaris at the altitude 14.5 degrees is about-3, 7 minutes or-222 sec.arc. This value is greater than astronomical refraction of the star that being closest to the Sun. Thus we can draw a conclusion that spacetime is false.

5.Astronomical data of 1973 solar eclipse in Mauritania (Africa).

On the 1973 solar eclipse in Mauritania (Africa), the altitude of maximum eclipse is about 83.5 degrees. The astronomical refraction is about -9 sec arc, and the terrestrial refraction/DIP is about -168 sec arc. See on the Figure 3.7 below.

1973 Eclipse - Mauritania (Africa) Lat : 17.644^0 N Long: 3.8672^0 W				
Event	Date	Time (UT)	Alt	Azi
Start of partial eclipse	1973/06/30	10.01.39	64.5^0	073.0^0
Maximum eclipse	1973/06/30	11.33.51.4	83.5^0	030.5^0
End of total eclipse	1973/06/30	13.04.57.1	71.2^0	290^0

Deviation of a starlight closest to the Sun (App.Lat Corr for 83.5^0)

Astronomical refraction	DIP/Terrestrial refraction
- 0,15 minutes = - 9 sec.arc.	- 2,8 minutes = -168 sec.arc.

Figure 3.7: 1973 eclipse data of Mauritania

Then value of the deflection of light is found -177 sec arc, that's more than 100 times greater than Einstein's prediction. So, we can draw a conclusion that Einstein's prediction really doesn't work. General relativity was totally wrong.

While Mauritania is located on the Northern hemisphere; the observers in Mauritania may be being able to see the North Star of Polaris during the maximum solar eclipse. Altitude of Polaris is the same with latitude of Mauitania, it about 17.64 degrees. The similar situation with the 1952 solar eclipse in Sudan, that's also the best opportunity to re-examine the validity of the curvature of spacetime.

Altitude of Polaris star is about 17.64 degrees; astronomical refraction is about -2,95 minutes or -177 sec.arc, it's more than 100 times greater than astronomical refraction of the star that being closest to the Sun 1.75 sec arc. Thus, the astronomical data of 1973 eclipse in Mauritania (West-Africa) proves Einstein's prediction does not work, and spacetime is false.

As we know that general relativity proof not depends only on the eclipse verification. But, it must be considered, first, Einstein himself dared to admit: "If a single one of the conclusions drawn from it proves wrong, it must be given up; to modify it without destroying the whole structure seems to be impossible." Second, the primary problem of general relativity is not the way to testing hypothesis, but the hypothesis itself. Einstein's hypothesis of general relativity is not valid and doesn't meet requirements of the scientific method. The consequence is general relativity can not be proven or tested in any way.

Have you ever heard the statements below?

1. "One century after its formulation, Einstein's general relativity has made remarkable predictions and turned out to be compatible with all experimental tests."; or

2. "Special and general theory of relativity are incredibly well tested and very accurate theories".

These statements actually are incorrect. The entire experiments with the result 'general relativity is correct' or 'Einstein was right again' actually are 'the experiment is made base on belief' or the case of 'knowing the result'

In fact, for more than 90 years the Nobel Committee statement in the year 1921 is still valid:"Without taking into account the value that will be accorded your relativity and gravitation theories after these are confirmed in the future".

To many, and to Einstein himself, this felt like a slap in the face.[8]

4. The Gravitational Redshift is False.

From the previous discussion two important points have been discussed, firstly that special relativity is inconsistent with general relativity. Secondly, the hypothesis of general relativity is invalid. Therefore, the term of gravitational redshirt raises many questions, how can it be true?

The website Einstein online tells us about gravitational redshift as follow:

One of the three classical tests for general relativity is the gravitational redshift of light or other forms of electromagnetic radiation. However, in contrast to the other two tests - the gravitational deflection of light and the relativistic perihelion shift -, you do not need general relativity to derive the correct prediction for the gravitational redshift. A combination of Newtonian gravity, a particle theory of light, and the weak equivalence principle (gravitating mass equals inertial mass) suffices. It is, therefore, perhaps best regarded as a test of that principle rather than as a test of general relativity.

The gravitational redshift was first measured on earth in 1960-65 by Pound, Rebka, and Snider at Harvard University, who examined gamma rays emitted and absorbed by atomic nuclei. The gravitational redshift of light coming to us from the sun has also been observed, but the accuracy is not very good because of gas motions on the solar surface: Whenever light is emitted by a moving source, there is a motion-dependent frequency shift called a Doppler shift, and in the case of the sun, the Doppler shifts due to the moving gas are somewhat larger than the gravitational redshift due to the light having to climb out of the field of the sun.

Unfortunately, Einstein's mistake with the Doppler Effect. The Doppler effect allows us to very accurately measure the relative motion between a source and observer but by its very nature it does not allow us to determine the absolute motion of either. Einstein's mistake was to conclude from the Doppler effect that motion itself was intrinsically relative and not just hidden from the view of observers. Einstein failed to believe in a fixed frame that connected all forms of motion. However, with a more careful look at the Doppler effect, one must conclude that a common absolute motion for all photons must exist.

Photons provide the ultimate example of absolute motion since the evidence virtually proves that all photons move at exactly C within the same inertial reference frame. Einstein must have concluded from the Doppler effect that photons have no intrinsic wavelengths as they travel through space and that until they are measured there is no difference between a gamma-ray photon and a visible light photon.

For an extreme example of this, consider an observer traveling in a spaceship at a velocity of .999999 C from the earth to Alpha Centuari. Such extreme velocity would cause the observer to measure

Doppler shifts of between about Z = 1000 and Z = - 1000.

Gamma-ray photons coming from earth would be measured aboard the ship to be visible light photons and visible light photons from Alpha Centuari would be measured as Gamma-ray photons. The temperature of the 2.7° CBR in the direction of Alpha Centauri would be measured to be 2700°K and only .0027°K in the direction of Earth. Even at this high velocity the 2.7CBR would maintain the proper blackbody intensity and distribution curves for the temperatures measured. In the Living Universe, a photon must have an absolute intrinsic wavelength as it travels at C through space relative to the photon rest frame.

That photon is then Doppler shifted to some other virtual wavelength by the absolute motion of an observer. [9]

Redshift caused by refraction, not gravity

It is quite clear that deflection of starlight caused by refraction, not gravity. The term of gravitational redshift as we know is the term in the frame deflection of starlight by the Sun or light bending by gravity field of massive object.

Just for emphasize, the lights of objects in the sky reaching the earth have passed through layers of the terrestrial atmosphere, known as having different air density. Closer to the earth surface, the air is denser compared to the density of the air layer above it. The density is getting looser or weaker when it is getting higher.

If a ray of lights passes through from one medium to the others with different densities, such ray of lights will be reflected. The magnitude of refraction angle depends of density of its medium. That is why there are index of refraction. A ray of lights is passed though water, the said ray of lights will be reflected closer to the normal. A ray of lights is not reflected if its track is at the same direction with the normal.

The difference between air density and water density is sufficiently big or in a sudden, therefore the light track in the air and in water looks like a broken line. It is completely different from the light track at the earth atmosphere. The air density at the layers of earth atmosphere changes gradually and regularly. This causes the light refraction in the form of a curve. And the effect of such curve, the apparent position of a star (celestial bodies: Sun, Stars, Planets) will always look higher than its true position.

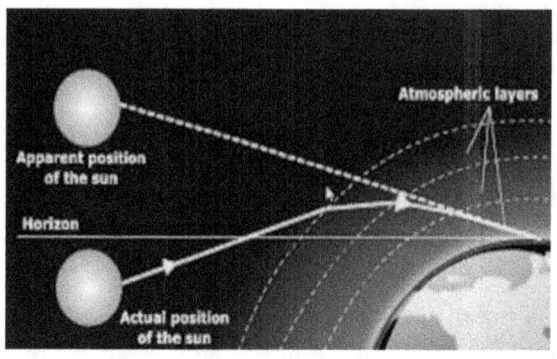

Figure 4.1:Apparent position of the Sun is always looks higher than its true position.(meritnationcom).

The Doppler effect or the Doppler shift is the change in frequency of a wavelength for an observer moving relative to its source. It is named after the Austrian physicist Christian Doppler, who proposed it in 1842 in Prague. It is commonly heard when a vehicle sounding a siren or horn approaches, passes, and recedes from an observer. Compared to the emitted frequency, the received frequency is higher during the approach, identical at the instant of passing by, and lower during the recession.

But the change in frequency of wavelength also related to temperature of its source. As the temperature increases, the frequency increase and the wavelength decrease. As tne wavelength increases, the frequency decreases, and as wave's wavelength decreases, the frequency increases. When electromagnetic energy is released as the energy level increases, the wavelength decreases and frequency decreases.

When the celestial bodies become reddering as temperature increase, they emit a red glow that we see as redshift. . If temperature continues to rise, reddering turn to orange, then the yellow, and then the blue that we see as a blueshift. The blue then turn to white. The amount of radiant energy given off by celestial bodies varied with wavelength and temperature.

This reddering is not related to the famous Doppler shift since the observer is not in motion relative to the body emitting the light signal.

Therefore, the term gravitational redshift can be misleading. No evidence redshift caused by gravity. A galaxys' redshift as Doppler effect actually is assumption that commonly accepted at that time in the life time of Einstein and other scientists. But in the views of the modern astronomy, a galaxys' redshift isn't Doppler. It is not cause by gravity, but by refraction..

Several ways can be conceived to explain this quantization. As noted earlier, a galaxys' redshift may not be a Doppler shift, it is the currently commonly accepted interpretation of the red shift, but there can be and are other interpretations. A galaxys' redshift may be a fundamental property of the galaxy. Each may have a specific state governed by laws, analogues to those in quantum mechanics that specify which energy states atoms may occupy. Since there is relatively little blurring on the quantization between galaxies, any real motions would have to be small in this model. Galaxies would not move away from one another; the universe would be static instead of expanding.[10]

The refraction-based theory is also able to explain gravitational red/blue shift. Also, black-hole, gravitational-lensing and space-time too are considered in the new perspective.

In view of the uncertainty & unavailability of information/data regarding refractive-index of atmospheric-medium and its variations; a rather semi-empirical approach, for the alternative explanation for bending of light near a star and gravitational red/blue shift etc., is appropriate and is described in the paper as follows.[11]

The limit of what space telescope can see

The most powerful telescope in history will never see the farthest galaxy. For example, the limit of Hubble Space Telescope, the farther out we look in the Universe, the redder any object's light will appear.

The first Hubble Space Telescope was launched into low Earth orbit in 1990. Hubble space telescope orbits at the altitude is about 600 Km above the Earth's surface. In other words, this telescope orbits in the thermosphere, not in outer space. The altitude of thermosphere is about 80–640 Km, and above it is exosphere at the altitude of 640–9600 Km.

Figure 4.2: Hubble Space Telescope -Wikipedia

Observations of the galaxy using Hubble Space Telescope at orbits in thermosphere, in fact, very difficult to do; that's much easier to do from the Earth. It's because the temperature of thermosphere which changes.

In the lower thermosphere, temperatures rise rapidly with altitude. Somewhere above an altitude between 200 and 300 km (about 320 to 480 miles) the temperature stays pretty much constant across altitudes.

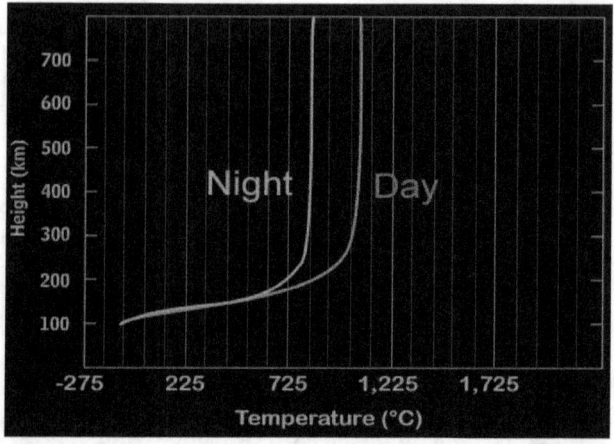

**Figure 4.3: Temperature in the Thermosphere-
windows2universe**

The hottest temperature in the thermosphere varies a lot between day and night and between the minimum and maximum levels of solar activity during the Sun's 11-year sunspot cycle.

For example, an image as result of Hubble Space Telescope is not much different from the image of the telescopes on Earth. See on the Figure 4.4 below.

Figure 4.4: New Hubble Space Telescope's image of Ursa Major's constellation, release date of observation 20 January 2014.

Former NASA Physicist Disputes Einstein's Relativity Theory

Dr. Edward Dowdye Jr. challenges Albert Einstein's fundamental theories as they are widely taught in schools today.

"I believe if Einstein were alive today, he would take advantage of the modern techniques and the modern instruments we have and he would wind up disproving his own theory," said Dr. Dowdye, a physicist and laser optics engineer who retired from NASA Goddard Space Flight Center. He is an independent researcher and founder of Pure Classical

Physics Research and he is a member of The American Physics Society.

Dowdye said he is part of a community of scientists who are questioning the relativity theory. He said he has gained the respect of a number of renowned physicists who agree with his stance. To name a few, Dr. Chandrasekhar Roychoudhuri, professor of physics at the University of Connecticut; Dr. Charles W. Lucas Jr., theoretical physicist and founder of Common Sense Science; and Dr. Edgar Kaucher, former member of the Institute for Applied Mathematics at the Karlsruher Institute for Technology, Germany.[12]

5. Is It True that Einstein's Relativity Incredibly Well Tested?

What happens if Gravity Probe B succeeded test general relativity?

According to NASA, the Gravity Probe B gyroscopes are the most perfect spheres ever made by humans. If these ping pong-sized balls of fused quartz and silicon were the size of the Earth, the elevation of the entire surface would vary by no more than 12 feet.

Gravity Probe B (GP-B) is a NASA physics mission to experimentally investigate Albert Einstein's 1916 general theory of relativity-his theory of gravity. GB-B uses four spherical gyroscopes and a telescope, housed in a satellite orbiting 642 km (400 mi) above the Earth, to measure in a new way, and with unprecedented accuracy, two extraordinary effects predicted by the general theory of relativity (the second having never before been directly measured):

1. The geodetic effect-the amount by which the Earth warps the local spacetime in which it resides.

2. The frame-dragging effect-the amount by which the rotating Earth drags its local spacetime around with it.

We are interested to know the relationships between spacetime and atmospheric medium; as we know in general relativity Einstein had ignored the atmospheric medium around a massive body, and then NASA tries to measure geodetic and frame-dragging effect in the Earth's atmosphere.

Earth's gravity pulls all the objects in atmosphere toward the Earth, and all the objects follow the rotation of the Earth. In the same way the satellite Gravity Probe B has orbit at the altitude 400 miles (642 km) above the Earth, remains in orbit at thermosphere/exosphere, and doesn't escape to space, it's because Earth's gravitational force. That is a facts; and in accordance with Newton's theory. No doubt about it.

On 4 May 2011, NASA announced the long-awaited results of Gravity Probe (GP-B), and a month later the results appeared in Phys. After more than 47 years and 750 million dollars, GP-B had succeeded in measuring the general relativistic geodetic and frame-dragging effects on orbiting gyroscopes. In this focus issue; CQG publishes a set of refereed papers that provide the complete details of the experiment, from design of the spacecraft to the final data analysis, thus bringing to a close an extraordinary chapter in experimental gravitation.

As already known, the satellite Gravity Probe B orbiting at the altitude 400 miles (642 km) above the Earth or in the exosphere; they measured geodetic effect and frame dragging in the atmospheric medium of the Earth. If Gravity Probe B had succeeded in measuring, it means Gravity Probe B had proven that spacetime is the same with atmospheric medium.

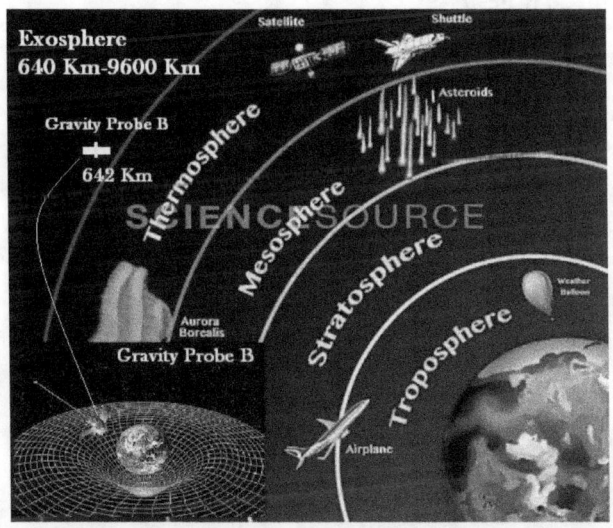

Figure 5.1: Gravity Probe B orbit in the atmosphere. Is spacetime the same with atmosphere?

If that is true, then the consequence is the confession that spacetime is the same with atmospheric medium. But, that is become deviate from each other. On one side, it says that test of general relativity succeeded. On the other hand, it is proven spacetime the same with atmospheric medium; this means general relativity is wrong.

What are the other consequences? First, if spacetime is the same with the atmospheric medium, then there is no black holes. In this case, Stephen

Hawking was correct as he states in Nature 'There are no black holes'

Most physicists foolhardy enough to write a paper claiming that "there are no black holes"—at least not in the sense we usually imagine—would probably be dismissed as cranks. But when the call to redefine these cosmic crunchers comes from Stephen Hawking, it's worth taking notice. In a paper posted online, the physicist, based at the University of Cambridge, UK, and one of the creators of modern black-hole theory, does away with the notion of an event horizon, the invisible boundary thought to shroud every black hole, beyond which nothing, not even light, can escape.

"There is no escape from a black hole in classical theory," Hawking told Nature. Quantum theory, however, "enables energy and information to escape from a black hole". A full explanation of the process, the physicist admits, would require a theory that successfully merges gravity with the other fundamental forces of nature. But that is a goal that has eluded physicists for nearly a century. "The correct treatment," Hawking says, "remain a mystery"[13]

Figure 5.2: **This is not a photo of a black hole. This is an artist's depiction of how matter might be ripped and radiate energy as it orbits the black hole's high-gravity center. Theoretically the black hole can not be seen in any way, though it uses the most advanced telescope.**

Einstein missing the distinction between time and clock.

The French philosopher Bergson was far more famous than Einstein in the first two decades of the 20th century. The reason most folks these days know Einstein's name but not Bergson's is, for Canales, an important story in itself. It's the story of how science seemed to become the last word on everything, even on a topic as subtle, slippery and difficult to pin down as time.

Bergson made it clear he had no problem with the mathematical logic of Einstein's theory or the data that supported it. But for Bergson, relativity was not a theory that addressed time on its most fundamental, philosophical level. Instead, he claimed, it was theory about clocks and their behavior. Bergson called Einstein out for missing the distinction.

The story from a prominent science writer, Lincoln Barnett, confirmed the French's philosopher:

According to Einstein. to describe any physical event involving motion, however, it is not enough simply to indicate position in space. It is necessary to state also how position changes in time. Thus to give an accurate picture of the operation of New York—Chicago express, one must mention not only that is goes from New York to Albany to Siracuse to Cleveland to Toledo to Chicago, but also the times at which it touches each of those points. his can be done either by means of a timetable or a visual chart.

In same way the flight of an aeroplane from New York to Los Angeles can best be pictured in a four-dimensional space-time continuum. The fact that the plane is at latitude x, longitude y, and altitude z, means nothing to the traffic manager of the airplane unless the time co-ordinate is also given. So time is fourth dimensional. And if one wishes to envisage the flight as a whole, as a physical reality, it can not be broken down into a series of disconnected take-offs, climbs, glides, and landing. Instead it must be thought of as continuum curve in a four-dimensional space-time continuum.

After explaining the flight of an aeroplane from New York to Los Angeles; actually, I expect Einstein explains the moving of stars and planets in the sky. Unfortunately, it seems Einstein has no idea on it. He doesn't know the fact, that positions of stars in the sky also can be figured at latitude x, longitude y, and altitude z. (Azimuth, Hour Angle, and Declination).

According to Einstein "since time is an impalpable quantity it is no possible to draw a picture or construct a model of a four-dimensional space-time continuum. But it can be imagined and it can be represented mathematically. Obviously the astronomers has to think of the Universe as space-time continuum." From this statement we know that the back ground of spacetime idea because he has no idea on the basic of astronomy; and not knowing that space and time has been applied in astronomy namely the celestial sphere coordinates system, for long time ago. Einstein admired to Riemann geometry and 4 D Minkovsky space, without considering the fact that Riemann geometry and Miskovsky space were not applied in astronomy.

Time is not the same as a clock, isn't it? What time is it now? Oh, at 9 o'clock.

What is term 'time'? A measurement system of nonspatial continuum in which events occur in apparently irreversible succession from the past through the present to the future.

What is a clock? An instrument other than a watch for measuring or indicating time, especially a mechanical or electronic device having a numbered dial and moving hands or a digital display.

Einstein said:"Gravity is nothing about force". It was written in about 1915, at that time has not been found the strength of gravity varies depending on location.

In fact, the gravity of Earth, which is denoted by g, refers to the acceleration that is imparted to objects due to the distribution of mass within the Earth. In SI units this acceleration is measured in metres per second squared (in symbols, $m/s2$ or $m \cdot s{-2}$) or equivalently in newtons per kilogram (N/kg or $N \cdot kg{-1}$). The precise strength of Earth's gravity varies depending on location. The nominal "average" value at the Earth's surface, known as standard gravity is, by definition, 9.80665 $m/s2$ (about 32.1740 ft/s2).

Einstein predicts gravitational waves (ripples in spacetime) in the year 1916 that is the same at that time has not been found gravity waves. Also, at that time has not been found that there is a place on earth that has a very strong force of gravity and a place has a gravitational anomaly.

In the Earth's atmosphere, gravity waves are a mechanism that produces the transfer of momentum from the troposphere to the stratosphere and mesosphere. Gravity waves are generated in the troposphere by frontal systems or by airflow over mountains. At first, waves propagate through the atmosphere without appreciable change in mean velocity. The effect of gravity waves in clouds can look like altostratus undulatus clouds, and are sometimes confused with them, but the formation mechanism is different.

A gravitational anomaly, for example, on the area of Bermuda Triangle. The Bermuda Triangle has a high gravitational field which exerts all particles towards it as it has a large number of volcanic openings under its sea surface. It may be come from the mysterious hexagonal clouds in Bermuda Triangle caused by 170mph 'air bombs' be behind centuries of bizarre disappearances. It is believed these deadly blasts of air that can flip over ships and bring planes crashing into the ocean could be behind the vanishing of at least 75 planes and hundreds of ships.

In fact, Einstein's theory of gravity cannot explain what the cause of wind and cloud. It can be explained by Newton's theory of gravity. Gravity compresses the Earth's atmosphere, it creates air pressure; the driving force of wind and creates clouds. Without gravity, there would be no atmosphere or air pressure and thus, no clouds and no wind. Its mean there would be no life at the Earth.

Gravitational Waves vs Gravity Waves

Gravitational waves are waves that vibrate through spacetime itself, as a result of gravitational forces as Einstein predicts in 1916. Gravitational waves are ripples in spacetime in accordance with Einstein's theory of gravity. According to LIGO, gravitational waves are ripples in spacetime caused by ultra-powerful cosmic explosions.

Gravity waves are waves driven by gravitational force. Gravity waves are waves in the Earth and atmosphere in accordance with Newton's theory of gravity.

After 100 years of theory and decades of experiments, astronomers have detected gravitational waves directly for the first time. The announcement, made Thursday by the Laser Interferometer Gravitational-Wave Observatory (LIGO) and accompanied by a paper in Physical Review Letters, describes a powerful signal that ultimately began with the merger of two black holes located 1.3 billion light-years away.

The finding not only confirms yet another aspect of Albert Einstein's theory of gravity, known as general relativity, but it also opens another avenue for researchers to observe and study the universe.

"We have detected gravitational waves. We did it," said David Reitze, executive director of LIGO, at a conference Thursday at the National Press Club in Washington D.C. "I am so pleased to be able to tell you that."

Before crashing together, the black holes were 36 and 29 times the Sun's mass. Afterward, the new combined black hole has only 62 solar masses, with the colossal difference — 5,000 supernovas' worth of energy — radiated away as gravitational waves.

These ripples open a new window on the universe, allowing astronomers to hear in the darkest regions of space where telescopes yield no information. Black holes, for instance, are infamously impossible to observe directly; they emit no light. But with gravitational waves, astronomers can probe the very heart of the singularities. They will discover black holes completely invisible to traditional observatories -and surely new surprises as well.

The idea of gravitational waves began with Einstein's theory of general relativity, and his realization that gravity was simply the warping of the fabric of space-time by massive objects. He figured out that massive moving objects would create ripples in this fabric, like a child's bounce on a trampoline. What's more, these ripples would propagate at the speed of light throughout the universe.

Figure 5.3: Einstein's original hand-written
(businessinsider.co.id)

Einstein predicts in 1916 that two celestial bodies in orbit will generate ripples in spacetime, not two black holes collide.

"the curator for Einstein's archives, Roni Grosz, points to one of the critical equations from Einstein's General Theory of Relativity, which predicts that *two celestial bodies in orbit will generate invisible ripples in spacetime that experts call gravitational waves.*"

Einstein didn't believe in black holes, he is the reluctant father of black hole. Two black holes collide nothing to do with Albert Einstein. It is something ironically, that the idea of gravitational waves began with Einstein's theory of general relativity, but Einstein himself didn't believe in black holes.

If gravitational waves exist, the speed of gravity/gravitational waves cannot be the same speed as the speed of light.

"The theory (special relativity) also tells us that nothing can travel faster than light. However, it was inconsistent with the Newtonian theory of gravity, which said that objects attracted each other with a force that depended on the distance between them. This meant that if one moved one of the objects, the force on the other one would change instantaneously. Or in other gravitational effects should travel with infinite velocity, instead of at or below the speed of light, as the special theory of relativity required."(Stephen Hawking)

Gravitational waves are not part of the electromagnetic waves. Gravitational waves and electromagnetic waves are quite different. Measuring the speed of gravity in September 2002 by Kopeikin, it is not measuring the speed of gravity, but the speed of light.

There is a claim from a scientist:

Open Letter to the Nobel Committee for Physics 2016, W.W. Engelhardt, JET, Max-Planck-Institut für Plasmaphysik. Abstract: The Nobel Committee is informed that according to Professor Karsten Danzmann (Albert Einstein Institut) the LIGO detectors are not calibrated as expected from the statement in the discovery paper: "The detector output is calibrated in strain by measuring its response to test mass motion induced by photon pressure from a modulated calibration laser beam [63]". The claim that gravitational waves have been detected is not substantiated experimentally, since direct calibration data, namely mirror displacement as a function of laser power moving the mirrors, are not published.[14].

Thus, gravitational wave is likely to be wrong. Using the law of logic we can says:"Between both of the waves driven by gravity, gravity waves of Newton's theory and gravitational waves of Einstein's theory; both cannot both be true, both can be wrong but only one can be right."

Time Dilation

In the theory of relativity, time dilation is a difference of elapsed time between two events as measured by observers either moving relative to each other or differently situated from a gravitational mass or masses. An accurate clock at rest with respect to one observer may be measured to tick at a different rate when compared to a second observer's own equally accurate clock. This effect arises neither from technical aspects of the clocks nor from the fact that signals need time to propagate, but from the nature of spacetime itself.

Time dilation of Albert Einstein's relativity, special and general relativity in these two circumstances can be summarized as follow: In special relativity (or, hypothetically far from all gravitational mass), clocks that are moving with respect to an inertial system of observation are measured to be running more slowly. This effect is described precisely by the Lorentz transformation. In general relativity, clocks at a position with lower gravitational potential – such as in closer proximity to a planet – are found to be running more slowly.

Shapiro Time Delay

According to general relativity, time dilation was caused by gravity. This was confirmed by an experiment conducted by Erwin Shapiro in 1964. The experiment was carried out with the radar signal that is directed to the planet Venus. The result of experiments show that there is a time difference between the radar signals travel from Earth to Venus and travel back from Venus to Earth. *It caused by gravity field of the Sun*, show the signal travel back from Venus to Earth slower approximately 200 micro seconds. The experiment was repeated several times, and the result is the same.

Unfortunately, Shapiro's experiment does not explain the effect of the refraction of light by layer of Earth's atmosphere. It seems that the light refraction was ignored. The signal travel back slower, it isn't caused by the Sun, but by layer of Earth's atmosphere or refraction.

Therefore, Shapiro time delay or gravitational time delay actually is false. Time delay of light was caused by refraction, not gravity.

Hafele-Keating Experiments

In October 1971, Joseph C. Hafele, a physicist, and Richard E. Keating, an astronomer, took four cesium-beam atomic clocks aboard commercial airliners. They flew twice around the world, first eastward, then westward, and compared the clocks against others that remained at the United States Naval Observatory. When reunited, the three sets of clocks were found to disagree with one another, and their differences were consistent with the predictions of special and general relativity.

General relativity predicts an additional effect, in which an increase in gravitational potential due to altitude speeds the clocks up. That is, clocks at higher altitude tick faster than clocks on Earth's surface. This effect has been confirmed in many tests of general relativity, such as the Pound–Rebka experiment and Gravity Probe A.

In the Hafele–Keating experiment, there was a slight increase in gravitational potential due to altitude that tended to speed the clocks back up. Since the aircraft flew at roughly the same altitude in both directions, this effect was approximately the same for the two planes, but nevertheless it caused a difference in comparison to the clocks on the ground.

In fact, that is true the clocks at higher altitude tick faster than clocks on Earth's surface. It is not caused by gravity, but by air density of atmosphere. Closer to the earth surface, the air is denser compared to the density of the air layer above it. The density is getting looser or weaker when it is getting higher.

It is has been known in traveling on an airplane. At higher altitude the density of atmosphere is getting looser or weaker, and less of friction on an airplane. Traveling in weaker density of atmosphere an airplane can move faster than in denser atmosphere.

6.GPS, VLBI, and Measuring the Speed of Gravity

GPS doesn't need Einstein's relativity.

There are at least 4 reasons GPS doesn't need Einstein's relativity

1.GPS measuring location, not measuring time.

It's about time dilation, both time dilation of special and general relativity. In the case of special relativity-time delay of light-it's clear that light can be bent by layer of atmosphere. Thus, speed of light is not constant. What can special relativity do to GPS? Nothing. Signals from GPS isn't sent back from the receiver on earth to the GPS.

University of London Professor Herbert Dingle showed why Special Relativity will always conflict with logic, no matter when we first learn it. According to the theory, if two observers are equipped with clocks, and one moves in relation to the other, the moving clock runs slower than the non-moving clock. But the Relativity principle itself (an integral part of the theory) makes the claim that if one thing is moving in a straight line in relation to another, either one is entitled to be regarded as moving. It follows that if there are two clocks, A and B, and one of them is moved, clock A runs slower than B, and clock B runs slower than A. Which is absurd.

2.Nothing about gravitational time dilation

In the case of general relativity, we know and no doubt: general relativity is not valid. We know that they did not hesitate to say: "One century after its formulation, Einstein's general relativity has made remarkable predictions and turned out to be compatible with all experimental tests."; or said that "special and general theory of relativity are incredibly well tested and very accurate theories." But, actually, these statement is nonsense.

About testing general relatiity via eclipse experiment using optical telescope; if it was difficult in 1995 , to see details of 1–2 seconds of arc, how much more difficult was it in the in 1919–1973 eclipse experiments? The difficulty of performing precise measurements of optical starlight deflection during an eclipse can be seen from the results of 1919, 1922, 1929, 1947, 1952, 1973 experiments.

Testing general relativity using VLBI (Very-long-baseline interferometry). VLBI is a type of astronomical interferometry used in radio astronomy. In VLBI a signal from an astronomical radio source, such as a quasar, is collected at multiple radio telescopes on Earth. The important things must be note, the purpose of VLBI is collecting signal in the form of invisible light, not to measure the altitude of a star and bending of light in the form of visible light. VLBI can not be use as a sextant in celestial navigation

In fact, for more than 90 years statement of Nobel Committe in the year 1921 is still valid:"Without taking into account the value that will be accorded your relativity and gravitation theories after these are confirmed in the future".

In his book A Brief History of Time, Stephen Hawking said: 'A confirmation of general relativity won the Nobel Prize!". Hawking seems to be saying this, as a hope or may be a joke.In fact, the Nobel Prize in Physics was awarded to Taylor and Hulse in 1993 for the discovery of a new type of pulsar; and nothing to do with general relativity.   Let's look at on the website nobelpize; 1993 Nobel Prize in Physics "for the discovery of a new type of pulsar, a discovery that has opened up new possibilities for the study of gravitation"

3.There are no official statements

How accurate is GPS? It depends. GPS satellites broadcast their signals in space with a certain accuracy, but what you receive depends on additional factors, including satellite geometry, signal blockage, atmospheric conditions, and receiver design features/quality. For example, GPS-enabled smartphones are typically accurate to within a 4.9 m (16 ft.) radius under open sky. However, their accuracy worsens near buildings, bridges, and trees.

From the above website dedicated by the USA government to the GPS, we know that GPS does not put forward anything about Einstein's relativity. In other words, there are no official statements.

4. Explanation from GPS's special consultant.

In the 1990's, Van Flandern worked as a special consultant to the Global Positioning System (GPS), a set of satellites whose atomic clocks allow ground observers to determine their position to within about a foot. Van Flandern goes on to discuss GPS clocks, which are often cited as being proof positive of Einstein's relativity. It may surprise you, but the GPS system doesn't actually use Einstein's field equations.

In fact, this paper by the U.S. Naval Observatory tells us that, while incorporating Einstein's equations into the system may slightly improve accuracy, the system itself doesn't rely on them at all. To quote the opening line of the paper, "The Operational Control System (OCS) of the Global Positioning System (GPS) does not include the rigorous transformations between coordinate systems that Einstein's general theory of relativity would seem to require."

At high altitude, where the GPS clocks orbit the Earth, it is known that the clocks run roughly 46,000 nanoseconds (one-billionth of a second) a day faster than at ground level, because the gravitational field is thinner 20,000 kilometers above the Earth. The orbiting clocks also pass through that field at a rate of three kilometers per second-their orbital speed. For that reason, they tick 7,000 nanoseconds a day slower than stationary clocks.

To offset these two effects, the GPS engineers reset the clock rates, slowing them down before launch by 39,000 nanoseconds a day. They then proceed to tick in orbit at the same rate as ground clocks, and the system "works." Ground observers can indeed pinpoint their position to a high degree of precision. In (Einstein) theory, however, it was expected that because the orbiting clocks all move rapidly and with varying speeds relative to any ground observer (who may be anywhere on the Earth's surface), and since in Einstein's theory the relevant speed is always speed relative to the observer, it was expected that continuously varying relativistic corrections would have to be made to clock rates. This in turn would have introduced an unworkable complexity into the GPS. But these corrections were not made. Yet "the system manages to work, even though they use no relativistic corrections after launch," Van Flandern said. "They have basically blown off Einstein".

Actually, the GPS satellites use classical (Newtonian) relativistic principles to work. These are the same relativistic principles that make sense in the radio astronomy which has been used since before the year of 1950.

VLBI Measurement

In the Letters to Nature, 28 February 1991, by D. S. Robertson, W. E. Carter & W. H. Dillinger: New measurement of solar gravitational deflection of radio signals using VLBI

"RADIO observations using very-long-baseline interferometry (VLBI) can measure the deflection of electromagnetic radiation by the Sun's gravitational field with an accuracy of better than 1 milliarcsecond, and can thus be used to test General Relativity. For an object at an angle a from the centre of the Sun, the expected deflection is1 (1 + gamma) (Ms/re)((l + cos alpha)/(l-cos alpha))1/2, where Ms is the mass of the Sun in geometrized units2 (1.477 times 105 cm), re is the distance from the Earth to the Sun in cm, and y is a parameter whose value is 1 if General Relativity is correct but which takes on different values in other theories of gravity. For gamma = 1, the deflection is 1,750 mas at the Sun's limb, 4 mas at alpha =90° and 0 at alpha = 180°. Our analysis of ten years of VLBI data, including observations of objects in the range 2.5° < a< 178°, yields an estimate gamma = 1.0002 with a formal standard error of 0.00096 and an estimated standard error of 0.002. This determination is comparable in accuracy and in good agreement with the determination from Mars–Viking time-delay measurements3".[15]

An astrophysicist Sabine Hossenfelder write: "By the 1990s, one didn't have to wait for solar eclipses any more. Data from radio sources, such as distant pulsars, measured by very long baseline interferometer (VLBI) could now be analyzed for the effect of light deflection. In VLBI, one measures the time delay by which wave fronts from radio sources arrive at distant detectors that might be distributed all over the globe. The long baseline together with a very exact timing of the signal's arrival allows one to then pinpoint very precisely where the object is located-or seems to be located. In 1991, Robertson, Carter & Dillinger confirmed to high accuracy the light deflection predicted by General Relativity by analyzing data from VLBI accumulated over 10 years. [16]

Considering the fact that radio observations using very-long-baseline interferometry (VLBI) was conducted by 'our analysis of ten years of VLBI data'; It means collecting data within ten years, then comparing and analyzing data. This method is not justified in scientifically of astronomy. To measure the angle of deviation of starlight, in astronomy applies direct observation and instantaneous.

Aside radio observations can not be made to direct observation and instantaneous, it seems to be illogical. How could it be happened in science? It is because the VLBI measurements can provide the data as if someone wants to get the result.

Wikipedia informs us some of the scientific results derived from VLBI include:

1. High resolution radio imaging of cosmic radio sources.

2. Imaging the surfaces of nearby stars at radio wavelengths (see also interferometry)—similar techniques have also been used to make infrared and optical images of stellar surfaces

3. Definition of the celestial reference frame

4. Motion of the Earth's tectonic plates

5. Regional deformation and local uplift or subsidence.

6. Variations in the Earth's orientation and length of day.

7.Maintenance of the terrestrial reference frame

8.Measurement of gravitational forces of the Sun and Moon on the Earth and the deep structure of the Earth

9.Improvement of atmospheric models

10.Measurement of the fundamental speed of gravity

11.The tracking of the Huygens probe as it passed through Titan's atmosphere, allowing wind velocity measurements.

Let's note point 1 and point 8; point 1 High resolution radio imaging of cosmic radio sources, can be used to test general relativity; its mean Einstein's gravity 'nothing about force' is true. And point 8 Measurement of gravitational forces of the Sun and Moon on the Earth and the deep structure of the Earth; its mean Newton's gravity is true.

In this case, they can prove that both theories are true. But it can not happens in science. For example, if someone says:"VLBI confirmed to high accuracy both Newton's gravity and Einstein's gravity", this is something logical fallacies. Both theories can be wrong, but both theories can not both be true; only one can be right.

As we know in the previous discussion, the main problem of general relativity is not test of hypothesis, but the hypothesis itself.

Measuring the Speed of Gravity

Isaac Newton thought the influence of gravity was instantaneous, but Einstein assumed it travelled at the speed of light and built this into his 1915 general theory of relativity.

Light-speed gravity means that if the Sun suddenly disappeared from the centre of the Solar System, the Earth would remain in orbit for about 8.3 minutes, the time it takes light to travel from the Sun to the Earth. Then, suddenly feeling no gravity, Earth would shoot off into space in a straight line.

But the assumption of light-speed gravity has come under pressure from brane world theories, which suggest there are extra spatial dimensions rolled up very small. Gravity could take a short cut through these extra dimensions and so appear to travel faster than the speed of light without violating the equations of general relativity.

But how can you measure the speed of gravity? One way would be to detect gravitational waves, little ripples in space-time that propagate out from accelerating masses. But no one has yet managed to do this.

Kopeikin found another way. He reworked the equations of general relativity to express the gravitational field of a moving body in terms of its mass, velocity and the speed of gravity. If you could measure the gravitational field of Jupiter, while knowing its mass and velocity, you could work out the speed of gravity.

Bending waves

The opportunity to do this arose in September 2002, when Jupiter passed in front of a quasar that emits bright radio waves. Fomalont and Kopeikin combined observations from a series of radio telescopes across the Earth to measure the apparent change in the quasar's position as the gravitational field of Jupiter bent the passing radio waves.

From that they worked out that gravity does move at the same speed as light. Their actual figure was 0.95 times light speed, but with a large error margin of plus or minus 0.25.

Their result, announced on Tuesday at a meeting of the American Astronomical Society meeting in Seattle, should help narrow down the possible number of extra dimensions and their sizes.[17]

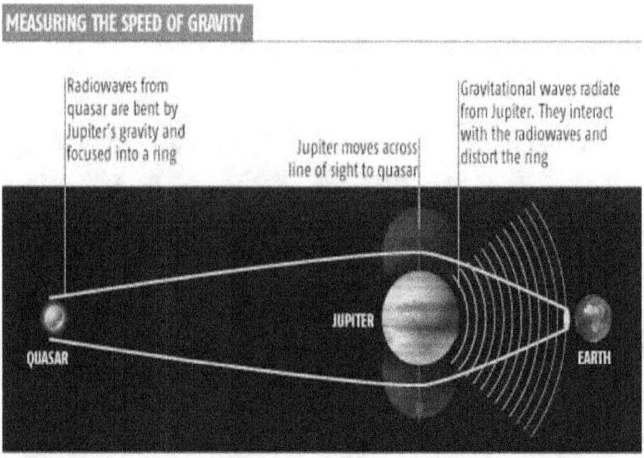

MEASURING THE SPEED OF GRAVITY

Radiowaves from quasar are bent by Jupiter's gravity and focused into a ring

Jupiter moves across line of sight to quasar

Gravitational waves radiate from Jupiter. They interact with the radiowaves and distort the ring

QUASAR   JUPITER   EARTH

Figure 6.1: Measuring the speed of gravity

Gravitational waves are not part of the electromagnetic waves. Gravitational waves and electromagnetic waves are quite different. Measuring the speed of gravity in September 2002 by Kopeikin—radiowaves from quasar are bent by Jupiter's gravity and focused into a ring-actually incorrect:

1.There is no evidence radiowaves from quasar are bent by Jupiter's gravity.

2.Radiowaves/electromagnetic waves are bent by refraction when travel throught layer of Earth's atmosphere.

3.In Kopeikin's experiment, actually they measuring is not the speed of gravity, but the propagation of radiowaves from quasar or they measuring the speed of light.

Several physicists, including Clifford M. Will and Steve Carlip, have criticized these claims on the grounds that they have allegedly misinterpreted the results of their measurements. Notably, prior to the actual transit, Hideki Asada in a paper to the Astrophysical Journal Letters theorized that the proposed experiment was essentially a roundabout confirmation of the speed of light instead of the speed of gravity. However, Kopeikin and Fomalont continue to vigorously argue their case and the means of presenting their result at the press-conference of AAS that was offered after the peer review of the results of the Jovian experiment had been done by the experts of the AAS scientific organizing committee.

In later publication by Kopeikin and Fomalont, which uses a bi-metric formalism that splits the space-time null cone in two – one for gravity and another one for light, the authors claimed that Asada's claim was theoretically unsound. The two null cones overlap in general relativity, which makes tracking the speed-of-gravity effects difficult and requires a special mathematical technique of gravitational retarded potentials, which was worked out by Kopeikin and co-authors but was never properly employed by Asada and/or the other critics.

Stuart Samuel also suggested that the experiment did not actually measure the speed of gravity because the effects were too small to have been measured. A response by Kopeikin and Fomalont challenges this opinion

Stuart Samuel, a participating scientist with the Theory Group of Berkeley Lab's Physics Division, in a paper published in Physical Review Letters, has demonstrated that an "ill-advised" assumption made in the earlier claim led to an unwarranted conclusion.

"Einstein may be correct about the speed of gravity but the experiment in question neither confirms nor refutes this," says Samuel. "In effect, the experiment was measuring effects associated with the propagation of light, not the speed of gravity."[18]

7.Closing

One of the fundamental concepts in astronomy, when taking observe for finding the angle of bending of star light in the sky, knowing the altitude of star is just as important as measuring the time of moving celestial bodies. We must know the altitude of star quite exactly if observations are to be of any use. The reason for this is that the entire of calculations depending on the altitude of star. Hypothesis and Einstein proposed test of general relativity via eclipse are closely related to astronomy, especially celestial navigation. For understanding the hypothesis and the test via eclipse; both are correct or incorrect, physics training is needed; but more importantly is celestial navigation training.

The other fundamental concepts in astronomy is how to compute deviation (bending, deflection) a certain star in the sky. It is important to note that astronomy applied direct observations and instantaneous; we must consider the moving celestial bodies; that's why timeliness is very important.

Unfortunately, we found the fact that Einstein had no idea on the basic astronomy, aside we found his two theories-special and general relativity-are .inconsistent.

In general relativity Einstein calculated the degree of deflection that should be observed and predicted that for the stars closest to the Sun the deviation would be about 1.75 seconds of an arc. This is without taking into account the altitude of the star/Sun as the object of observation. Thus, it's a big mistake because this prediction has no scientific meaning in astronomy.

Another fatal mistakes he wants measuring deflection of light by the Sun; but he proposed test measuring deflection of light by Earth's atmosphere; he had not realized about that. Ironically, this test is not scientifically correct and deeply wrong.

So, we can draw a conclusion that the main problem of general relativity is not test of hypothesis, but the hypothesis itself.

Hypothesis and Einstein proposed test of general relativity are closely related to astronomy, especially celestial navigation. For understanding that hypothesis and the test are not valid, physics training is needed; but more importantly is celestial navigation training. Unfortunately, physicists and astrophysicists are not trained to become experts in the field of celestial navigation. The navigators around the world will be easily to recognize the fatal flaws of these hypotheses and test.

Actually, general relativity can not be proven or tested in any way. No doubt, the entire tests that say 'general relativity is correct' really are the case of 'knowing the result they wanted to get'.

References

1. **Lincoln Barnett**, The Universe and Dr.Einstein, London, June 1949.

2.**The Speed** of Light and the Index of Refraction,rpi.edu.

3.**Why do** the planets go around the Sun?, spaceplace.nasa.gov

4.**Tom Van Flandern** Articles, http://www.ldolphin.org/vanFlandern/

5.**Miles Mathis**, The Perihelion Precession of Mercury, July 1, 2007.

6. **Gravitational deflection of light**, www.einstein-online.info/

7.**Einstein's** genius changed science's perception of gravity, www.sciencenews.org/

8. **Stuart Clark**, Why Einstein never received a Nobel prize for relativity, www.theguardian.com

9.**How Einstein was** Wrong about E=MC2,

www.circlon.com

10. **Red Shift** Riddles, http://cs.unc.edu/

11.**Prof. R. C. Gupta**, India, Bending of Light Near a Star and Gravitational Red/Blue Shift: Alternative Explanation Based on Refraction of Light.

12.**Dr. Edward Dowdye Jr**, Former NASA physicist disputes Einstein's relativity theory, June 16, 2014

13.**Stephen Hawking**, There are no black holes, Nature, 24 January 2014.

14.**W.W. Engelhardt**, Open Letter to the Nobel Committee for Physics 2016, June 2016

https://www.researchgate.net/publication/30458187 3_Open_Letter_to_the_Nobel_Committee_for_Phys ics_2016

15. **Letters to Nature**, 28 February 1991, by D. S. Robertson, W. E. Carter & W. H. Dillinger: New measurement of solar gravitational deflection of radio signals using VLBI

16.Sabine Hossenfelder, A wonderful 100th anniversary gift for Einstein

http://backreaction.blogspot.co.id)

17.Article, **First** speed of gravity measurement revealed, newscientist.com.

18.Article, Berkeley Lab Physicist Challenges Speed of Gravity Claim.

19.Bowditch, American Practical Navigator, Volume I - II, Defense Mapping Agency Hydrographic / Topographic Center, 1984.

20..John P.Budlong, Sky and Sextant, London, 1981

21..Stephen Hawking, A Brief History of Time,1988

22.Web:DiscoverMagazine

23.Web:NewScientist.

24.Web:BussinessInsider

25.Web:Wikipedia.

26.Web: arxiv.org/ftp/physics

About the Author

Capt (Ret) Gatot Soedarto, Navigator, was born in Tuban, East Java, he graduated from The Indonesia Naval Academy. Former Chief of The Indonesia Coast Guard's Fleet, lecturer at The Indonesia Naval Academy, lecturer at The Indonesia Naval Staff and Command College, lecturer at The Indonesia Armed's Command School.

The books of his works among others are: Computer Engineering (1981), Prevention and Coping with Fire Hazards (1983), Preventing the Environmental Destruction from Fire Hazard (1985), Sun Tzu and Naval Strategy (2012), Lessons of the Falklands War (2013), Eclipse 1919 and the General Relativity Theory (2014).

Twitter: @GatotSoedarto

Facebook:@Gatot S.Astari